This book belongs to:
<u>Linsey Imler</u>

May all of
Your Dreams
come True!

Kimberly Ken

THE RAINDROPS' ADVENTURE

To my children, Amber and Michael

Published by Kimberly Kerr Press
P.O. Box 63, Sewickley, PA 15143

No part of this book may be reproduced
or transmitted in any form or by any means,
electronic or mechanical, including photocopying,
recording or by any information system,
without the written permission of the Publisher.
For information regarding permission,
write to Kimberly Kerr Press, P.O. Box 63,
Sewickley, Pa 15143

ISBN: 0-9672073-0-4

Library of Congress Catalog Card Number: 99-94384

Copyright 1997 by Kimberly Michaud Kerr

All rights reserved.
Printed in the United States of America
April 1999

THE RAINDROPS' ADVENTURE

From Raindrops to Rainbows

Written and Illustrated by: Kimberly Kerr

KIMBERLY KERR PRESS

FIRST EDITION

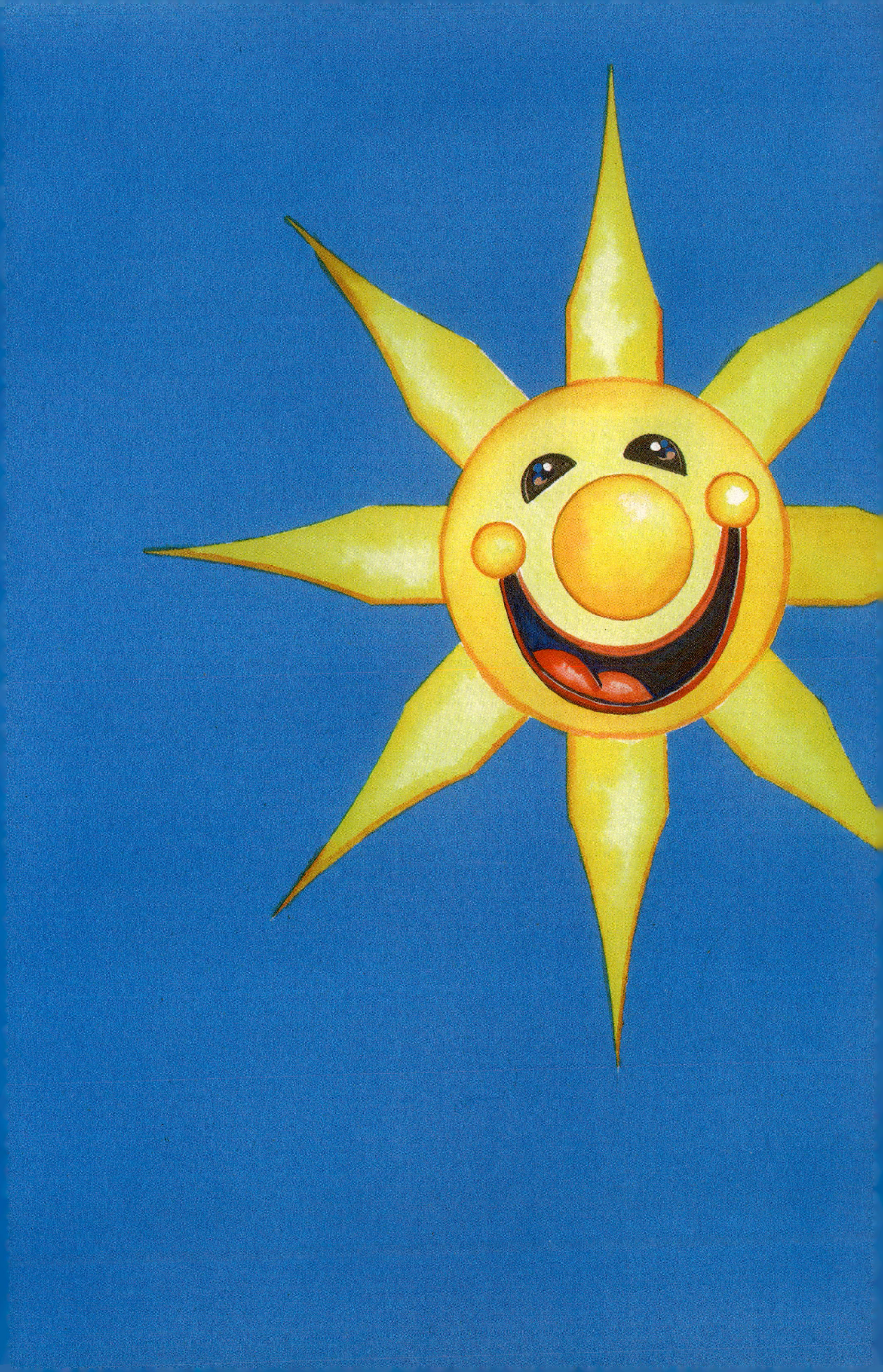

**See the Sun,
it shines so bright.
It gives off
a nice warm light.**

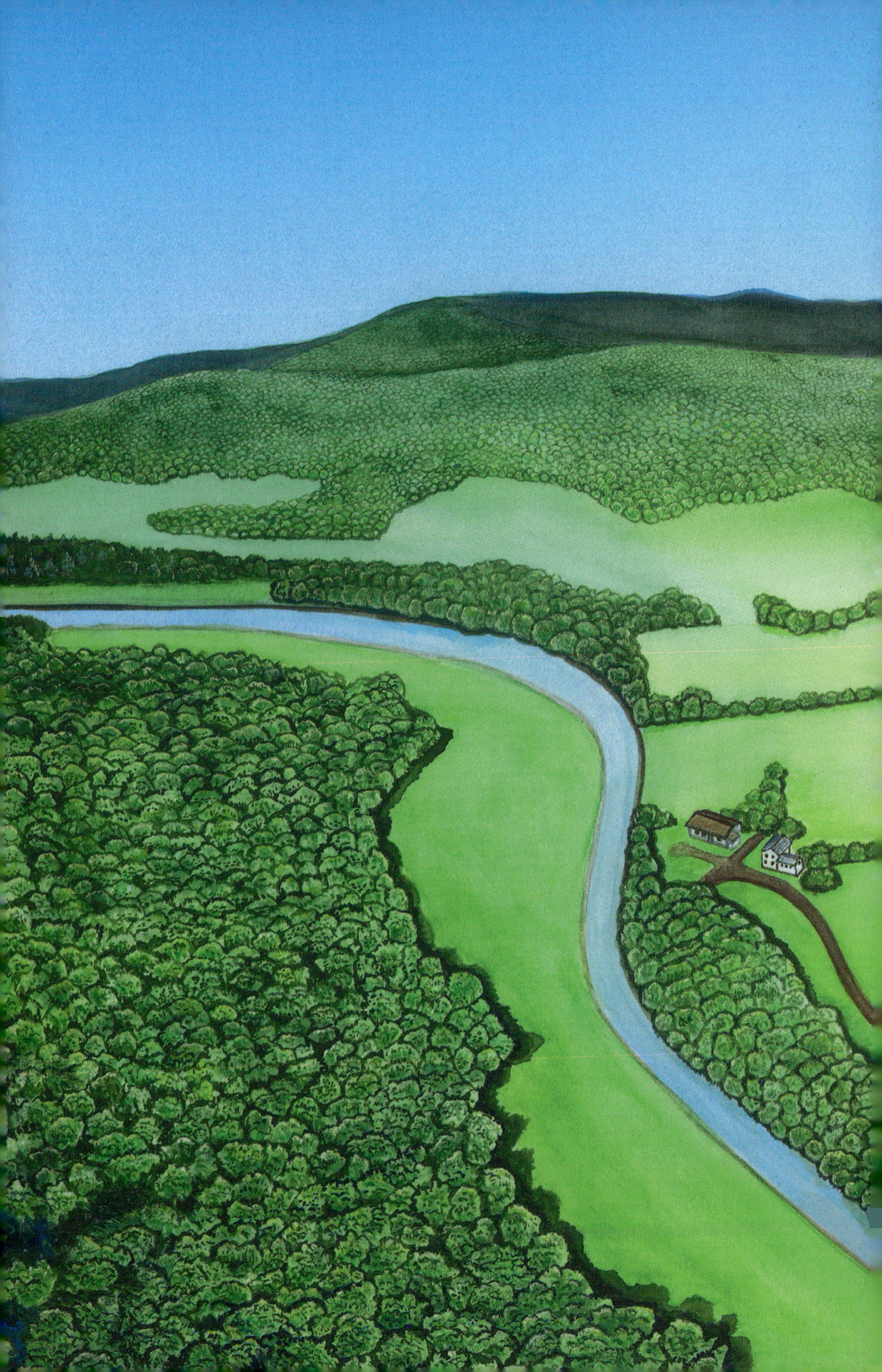

The Sun soaks up water
from the ground.
It pulls the water up
from all around.

The water vapor is pulled
up to the sky.
It's just pulled up.
Did you ever wonder why?

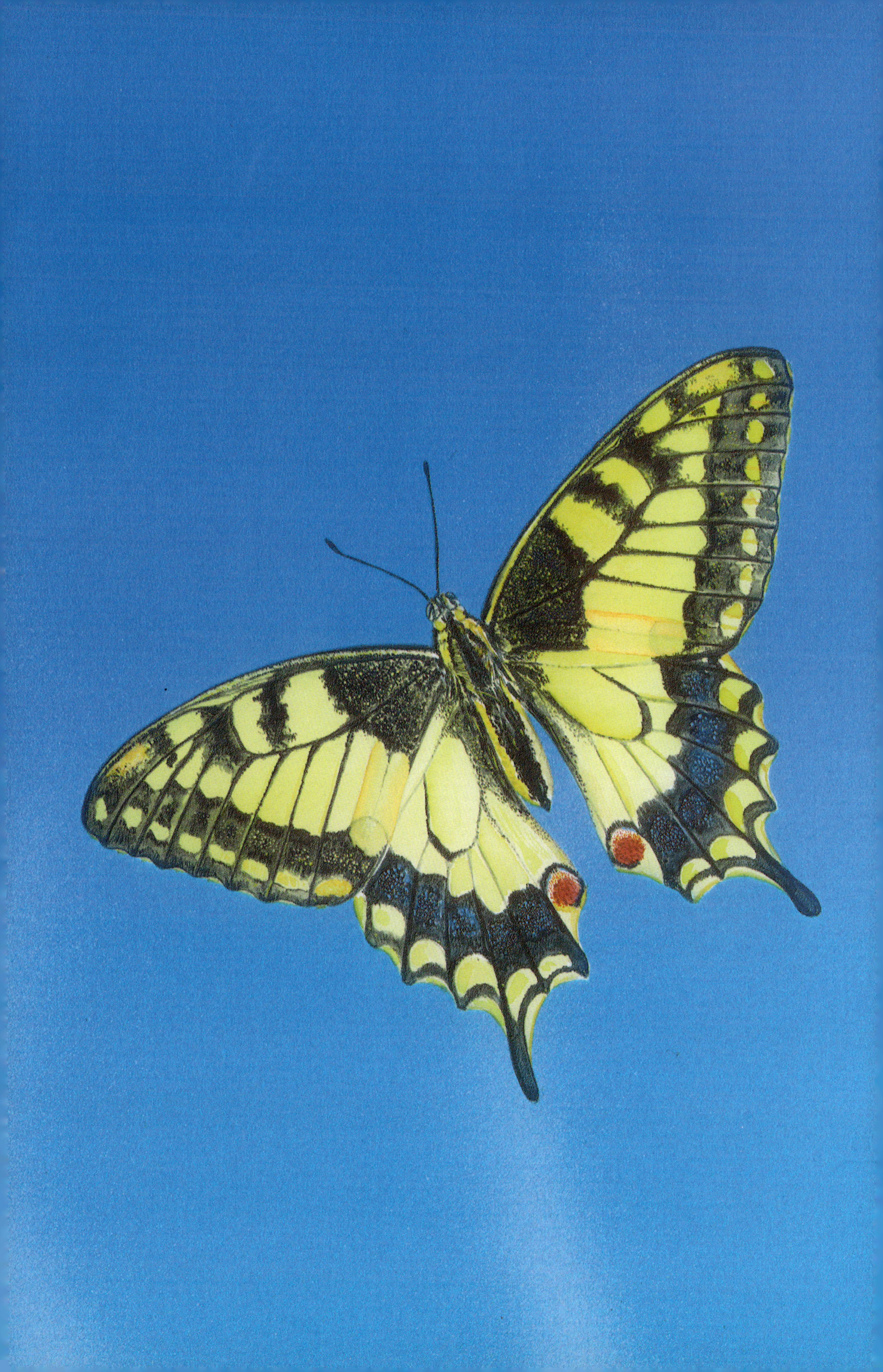

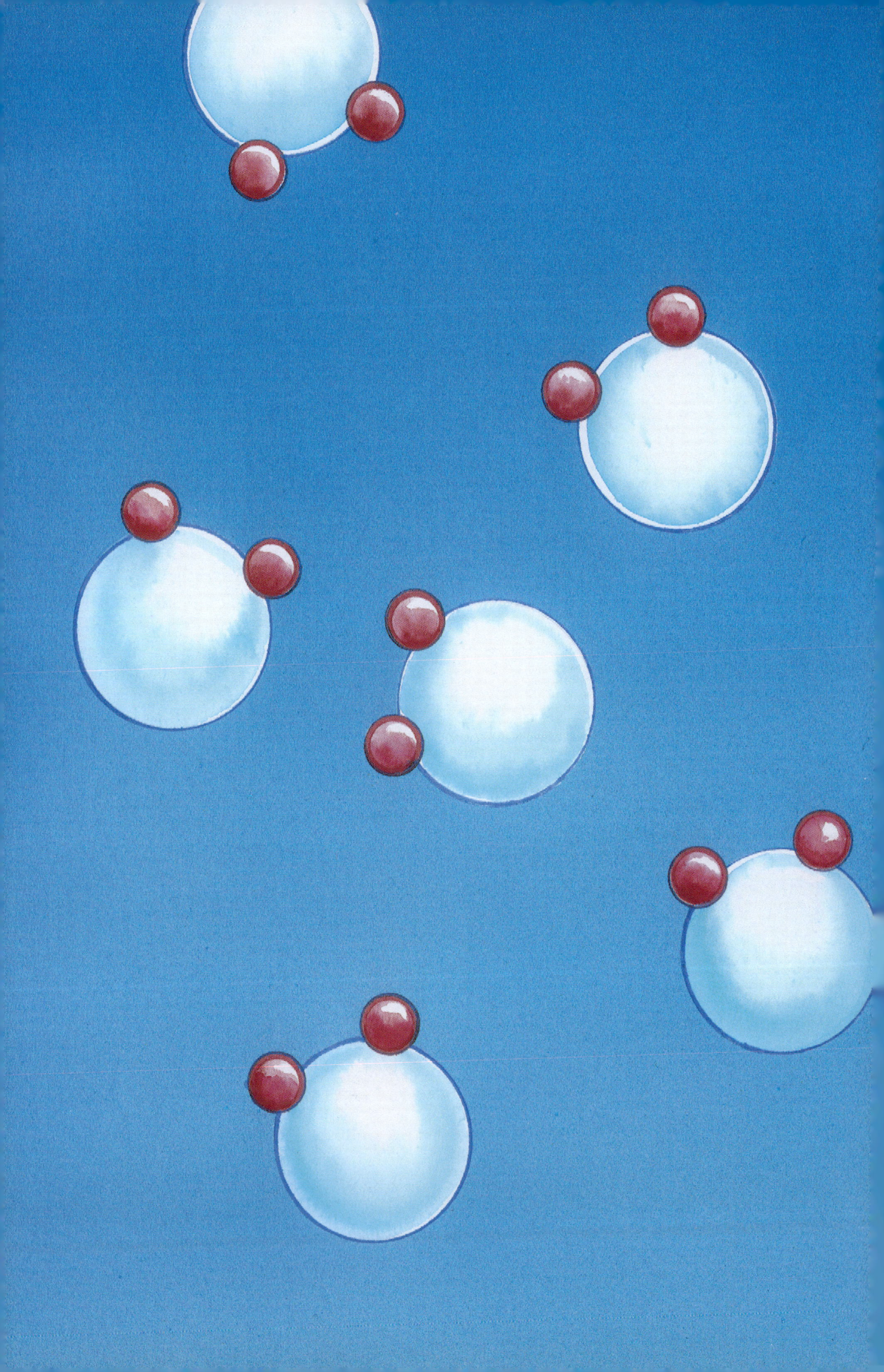

**This process
is called evaporation.
It's a really
neat sensation.
The higher
the molecules fly,
the cooler the air
is in the sky.**

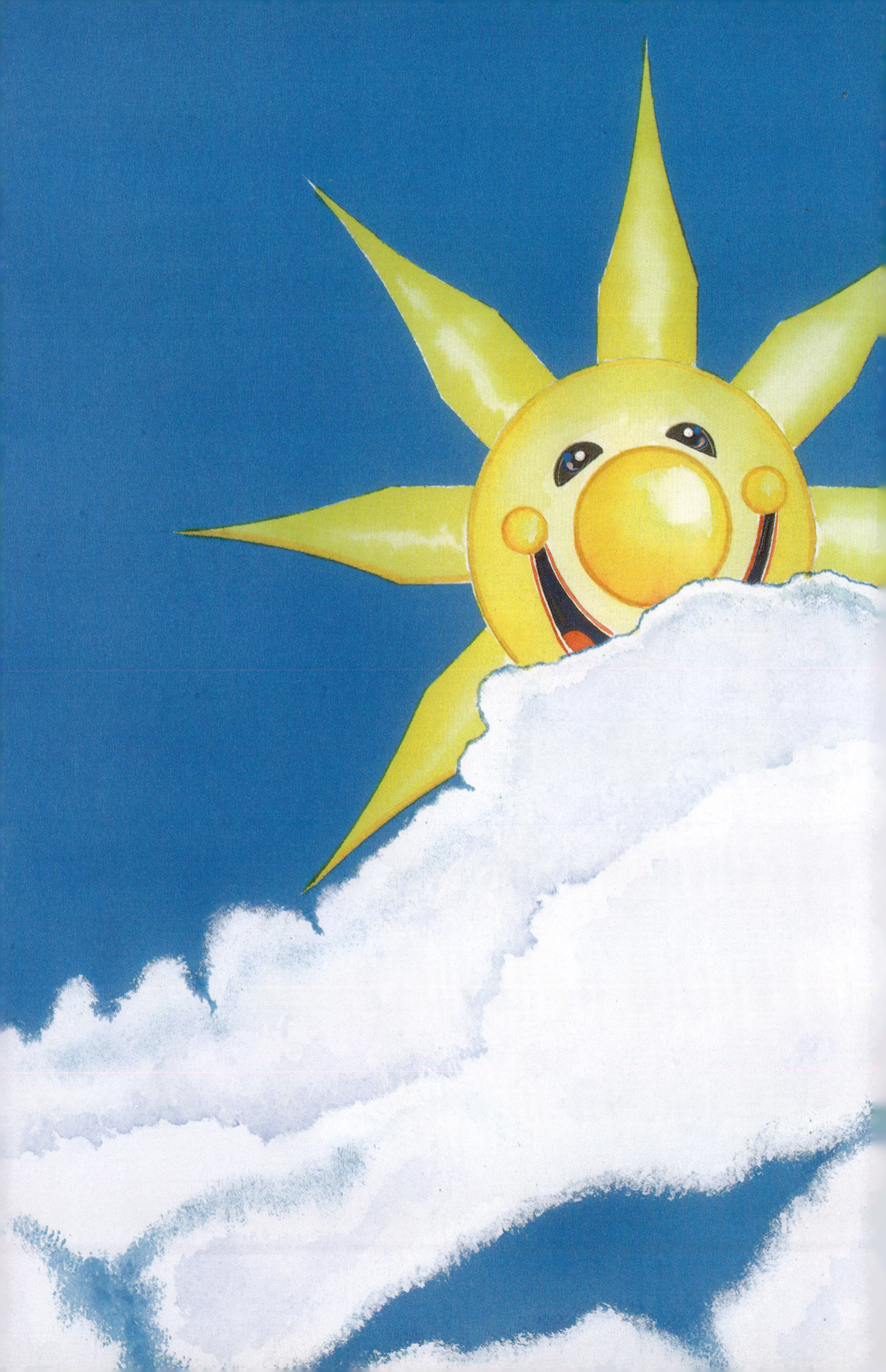

**Soon condensation
will happen up in the sky.
A cloud will start to form
right before your eye.**

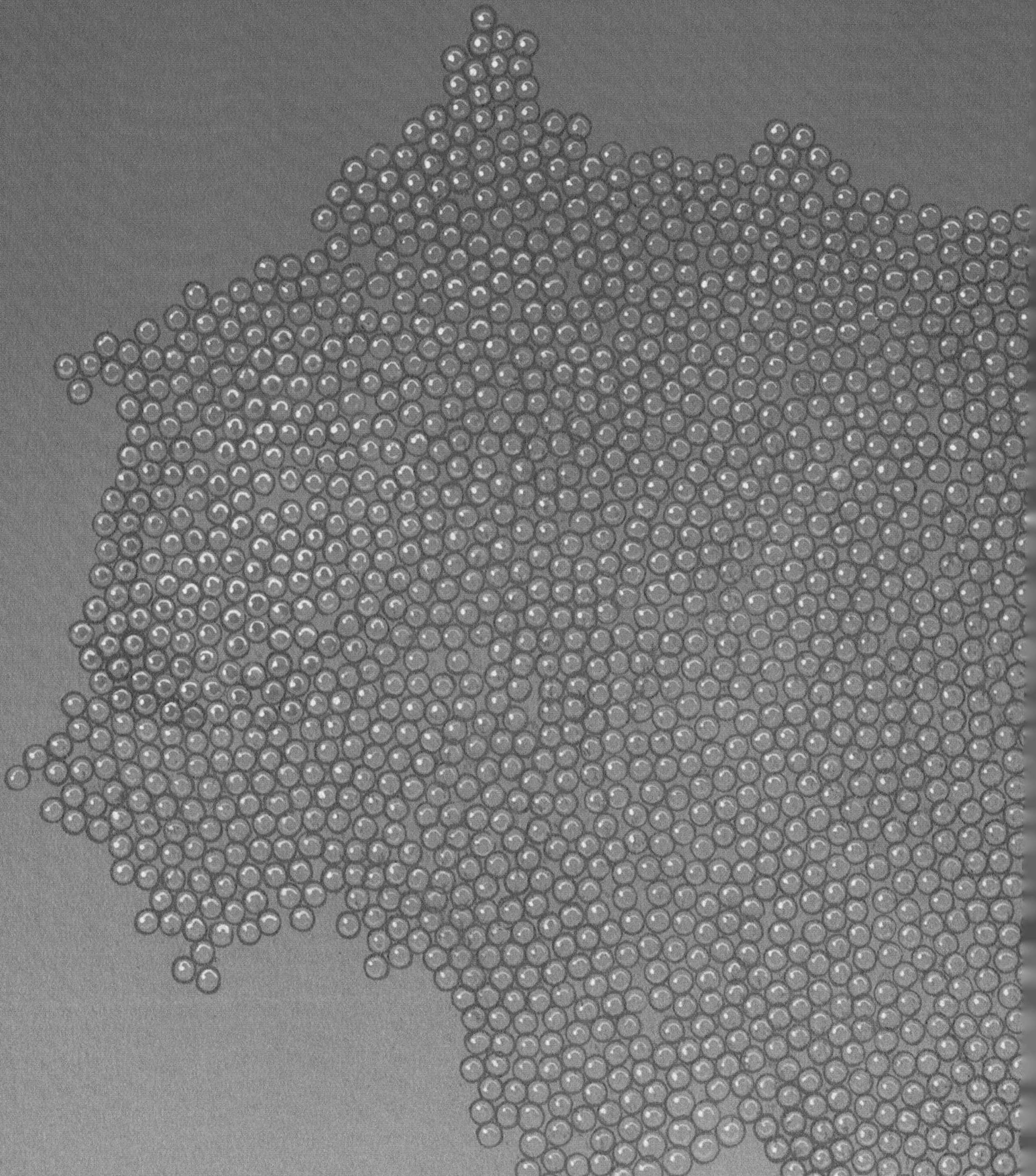

**The water vapors
will condense together.
Teeny tiny drops
of water will form.
They will condense together
to look like a great big swarm.**

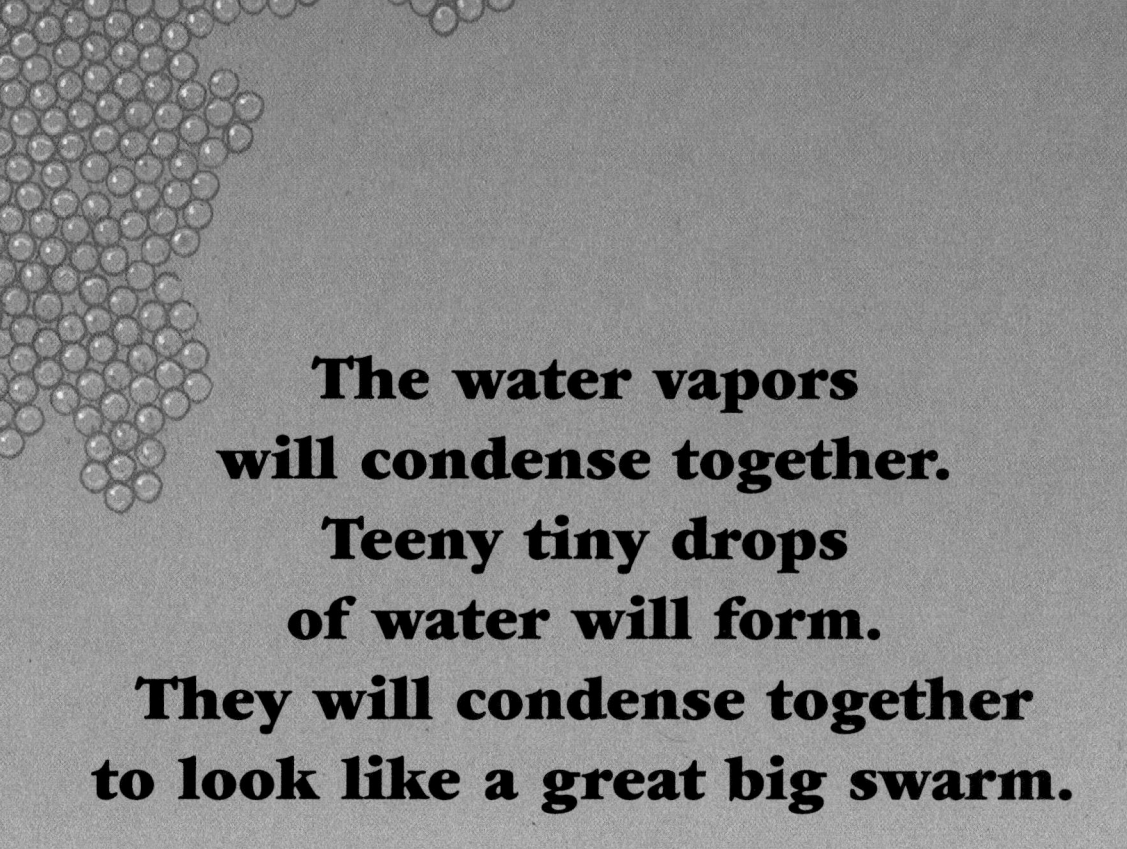

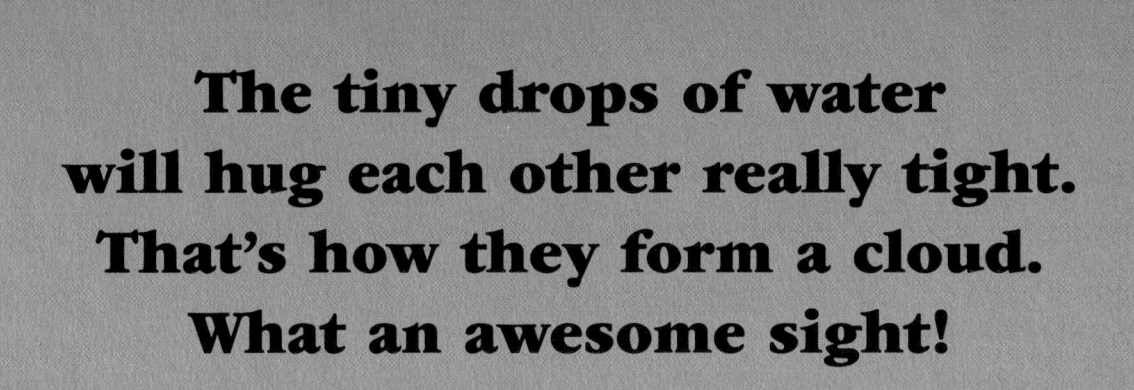

**The tiny drops of water
will hug each other really tight.
That's how they form a cloud.
What an awesome sight!**

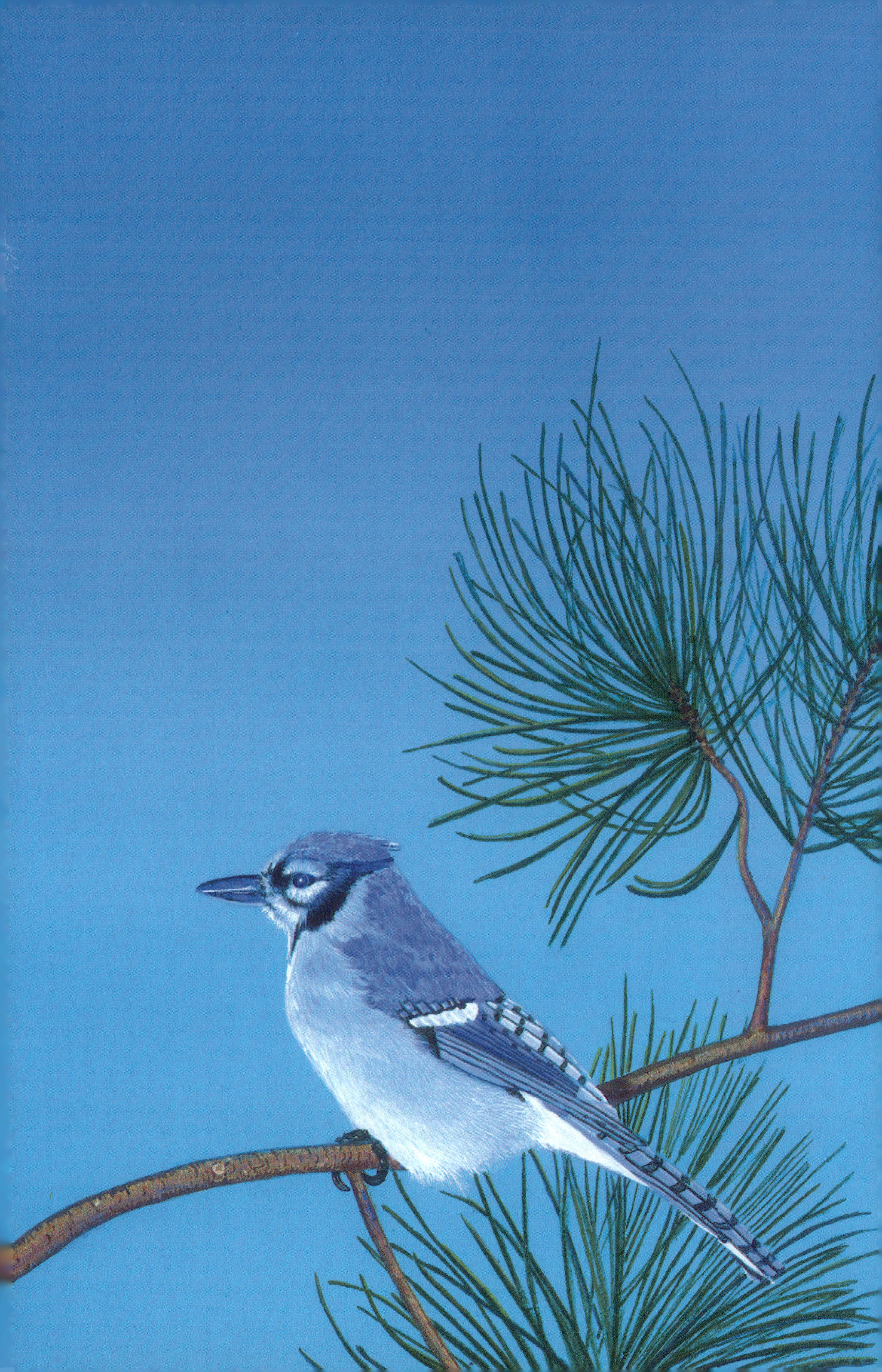

REMEMBER:

**To condense
is short for condensation.
The water vapor
turns to a liquid,
the opposite
of evaporation.**

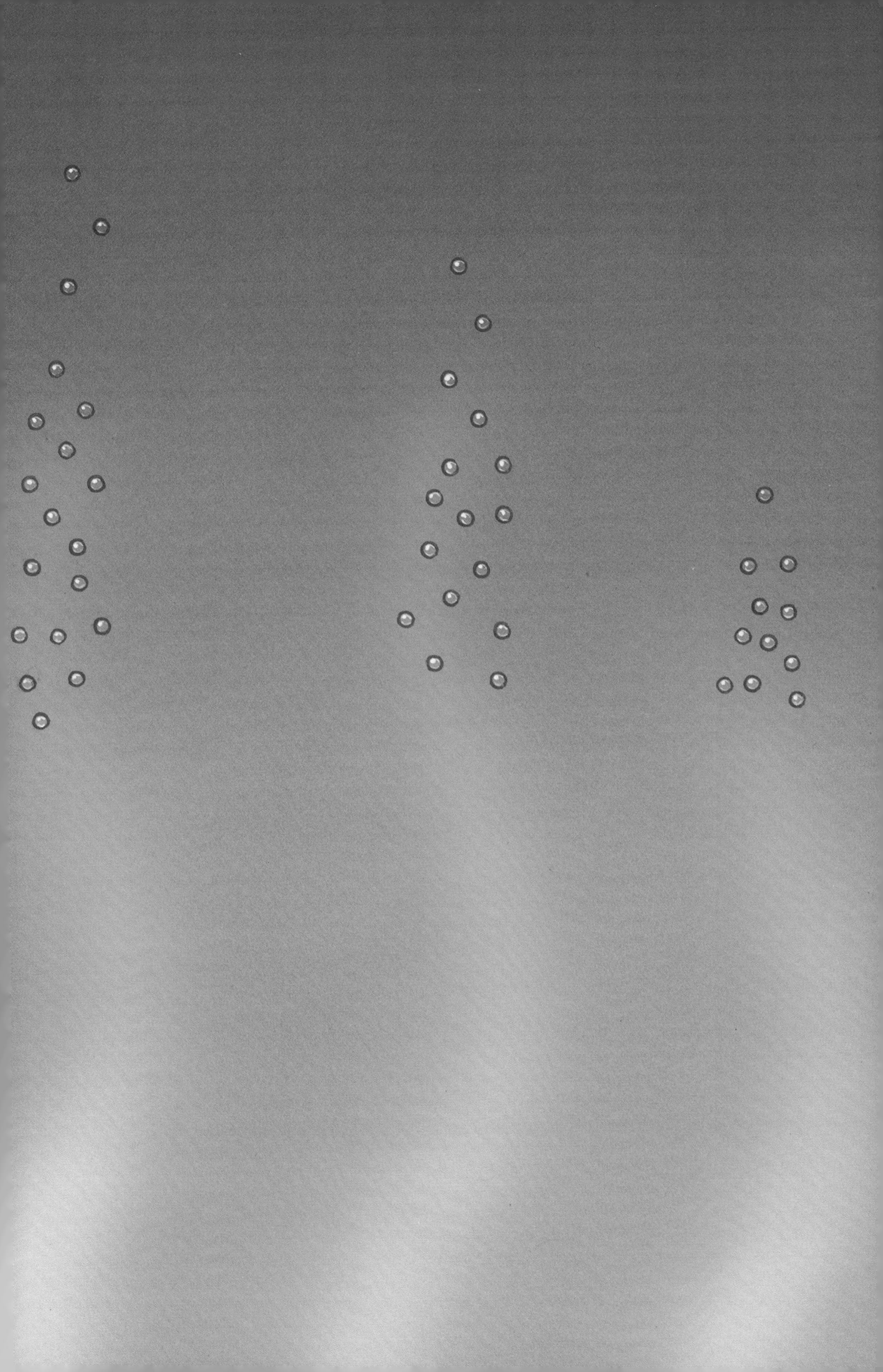

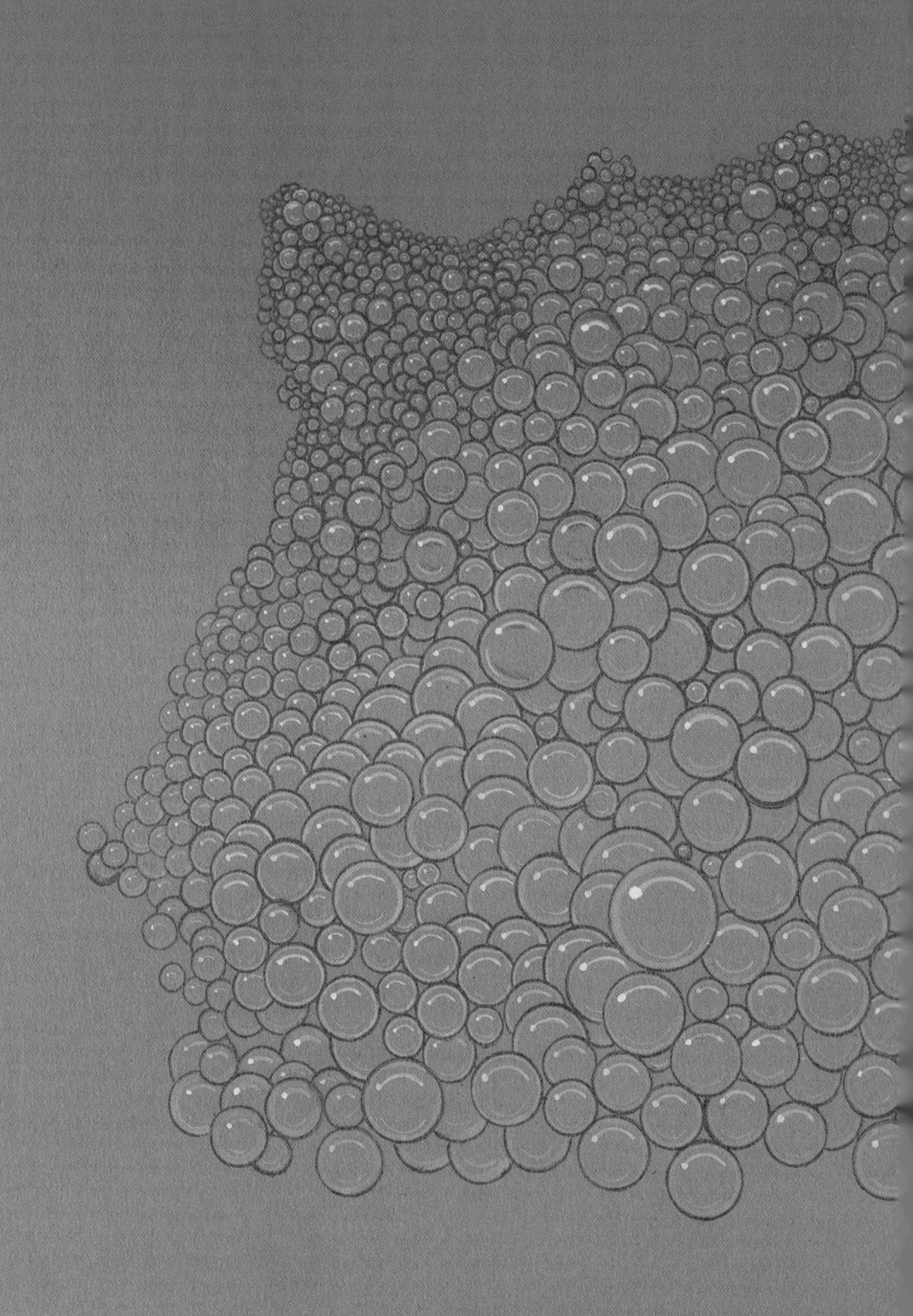

The cloud thinks
he is really clever,
but he can't hold
onto those drops forever.
They get bigger
and heavier, you see.
Then slowly, but surely,
they are set free.

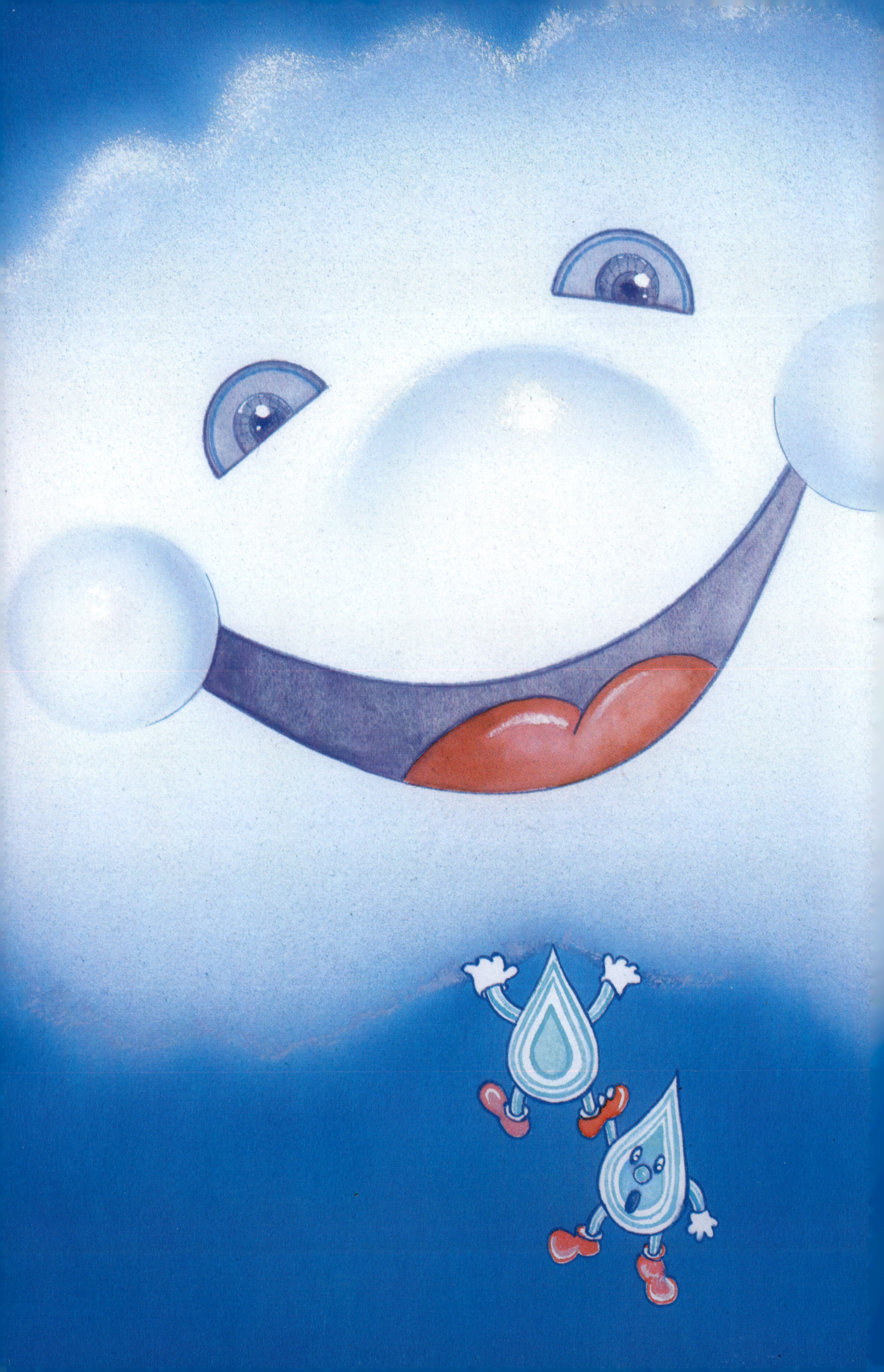

He begins to let go
one by one.
Wow! Raindrops have formed,
isn't this fun?

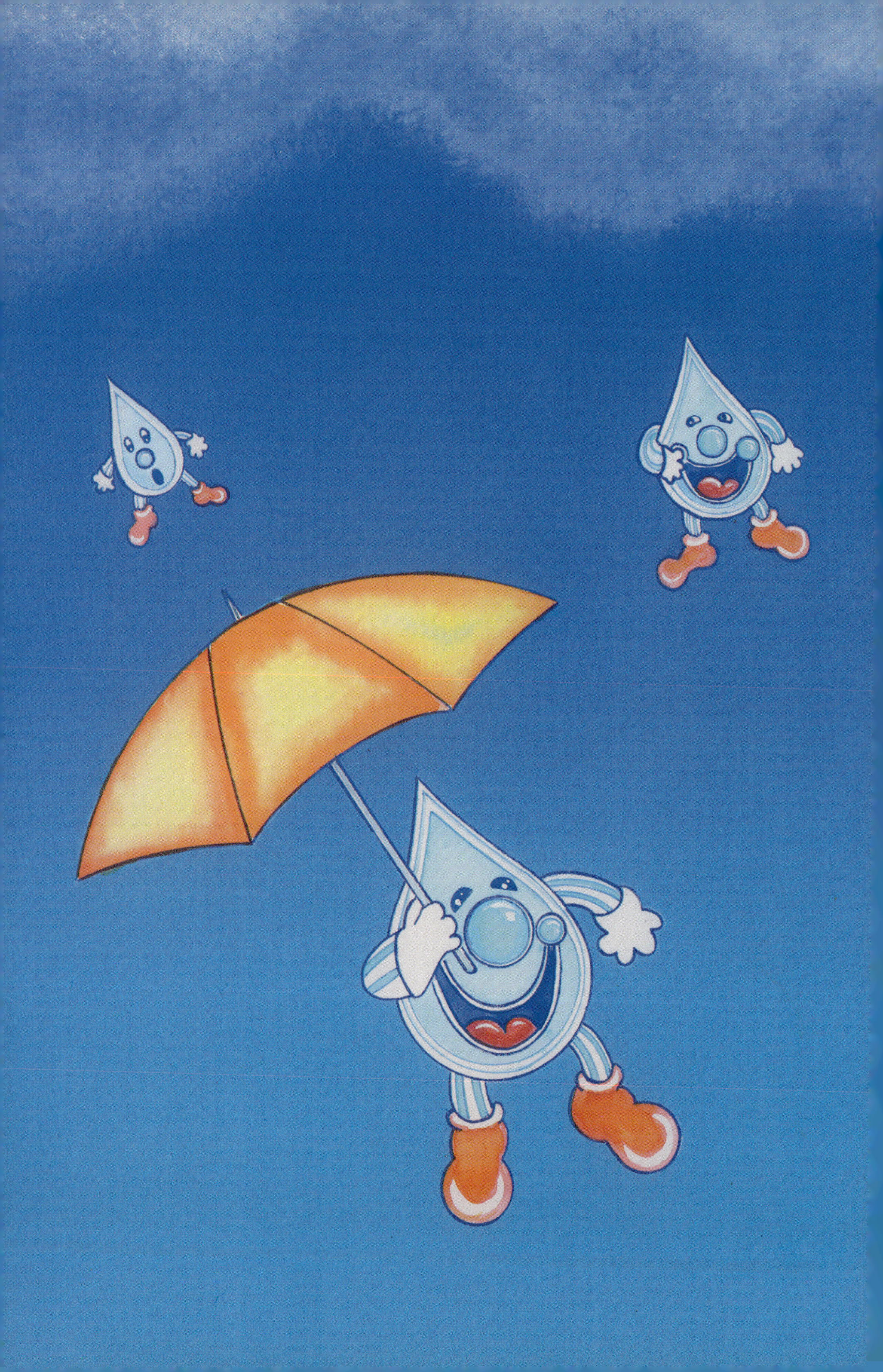

First one drop, then two,
three, then four.
Before you know it,
it begins to pour.
The raindrops laugh
and giggle and play.
They carry on
in the most unusual way.

Sometimes the sunlight shines through the raindrops forming a rainbow. How could this happen? Do you really want to know?

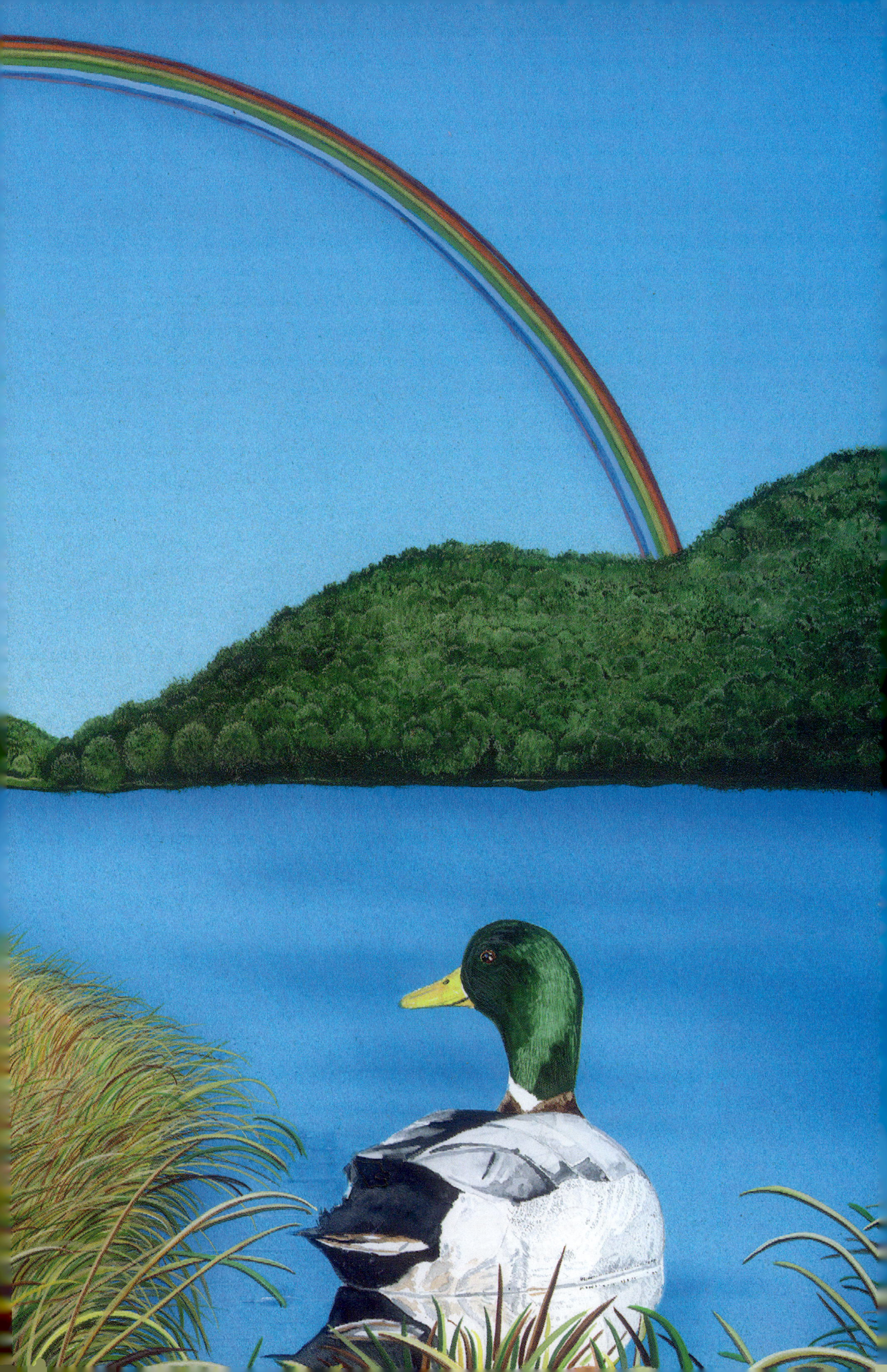

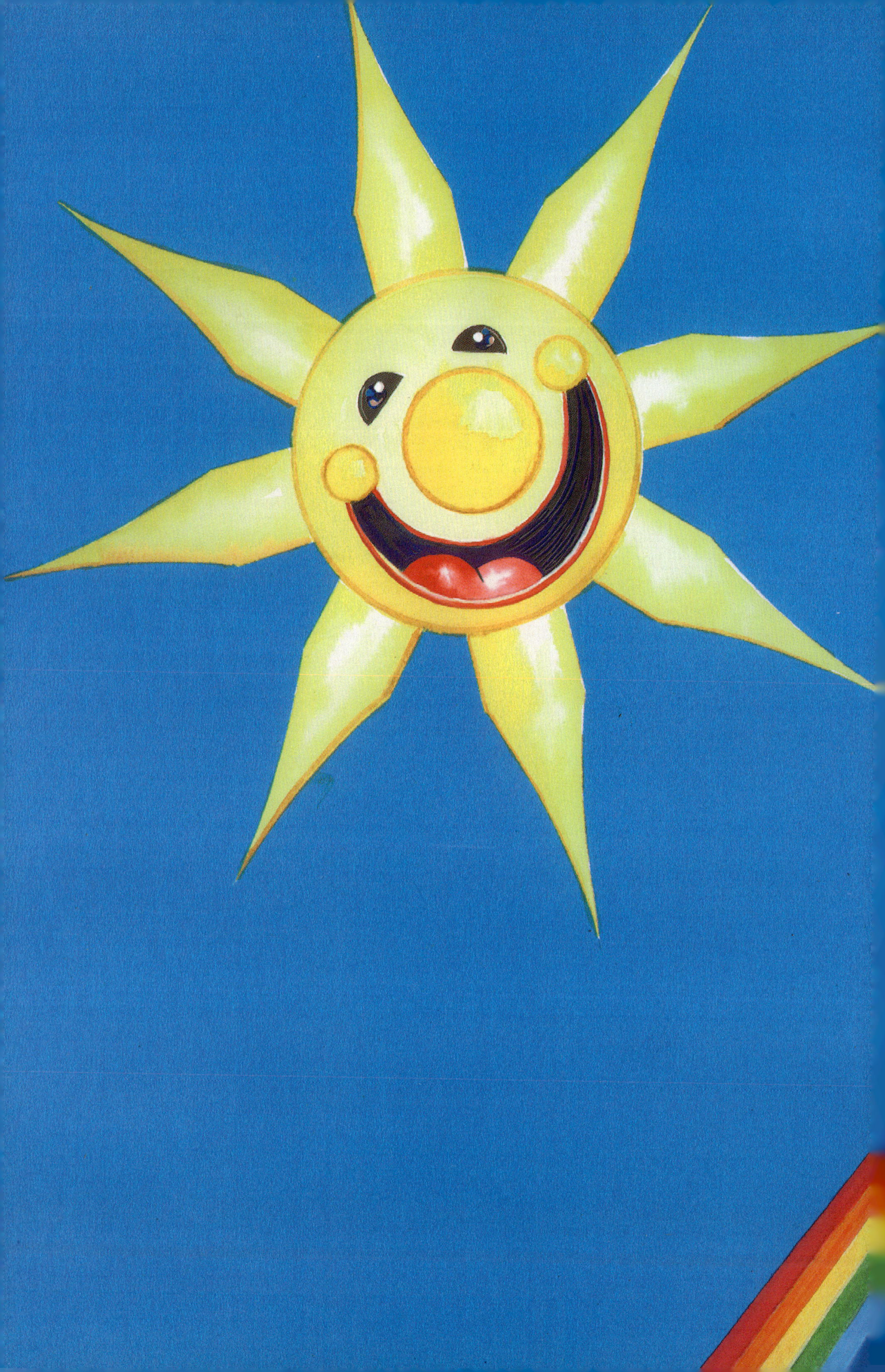

As the sunlight passes
through each drop,
the light is bent,
or refracted.
It doesn't want to stop.
It separates out into
the spectrum of colors
for you to see.
Red, orange, yellow, green,
blue, indigo, and violet.
It is a beautiful
sight to see.

There are four changing seasons
that come and go each year.
Just like the little raindrop
they again will soon reappear.

**Spring, Summer,
Fall, and Winter seasons,
the rain comes
for lots of reasons.**

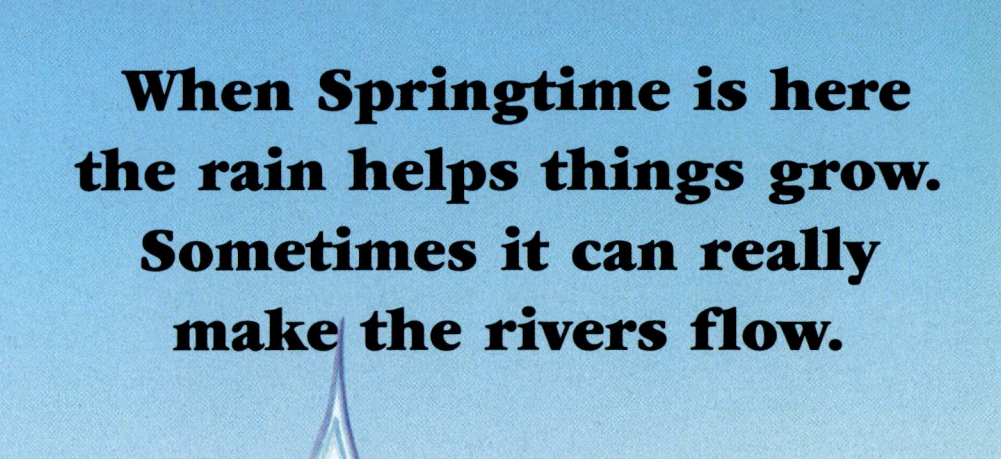

**When Springtime is here
the rain helps things grow.
Sometimes it can really
make the rivers flow.**

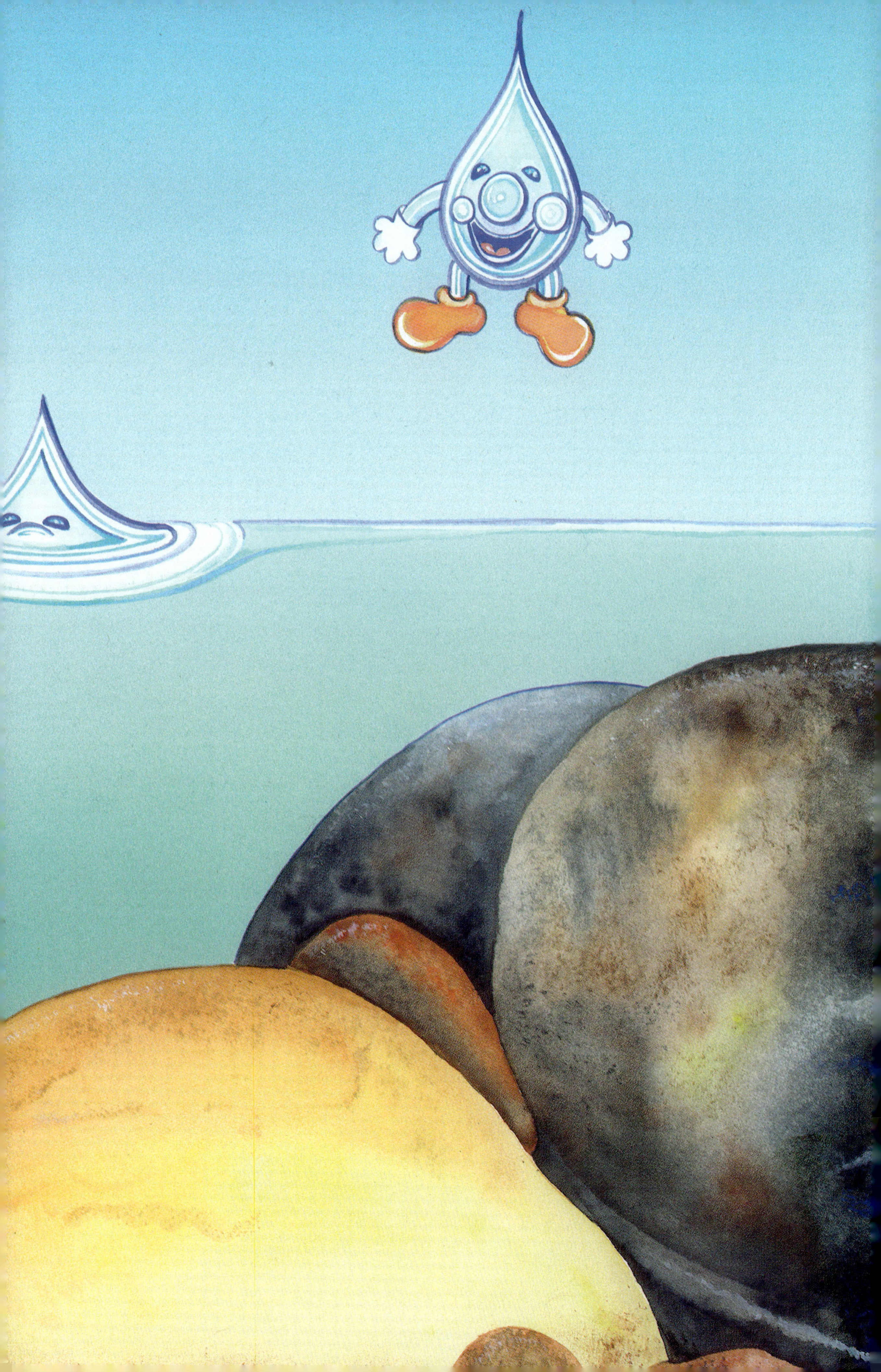

Summer nights can sometimes be frightening. That's when rain is mixed with thunder and lightning.

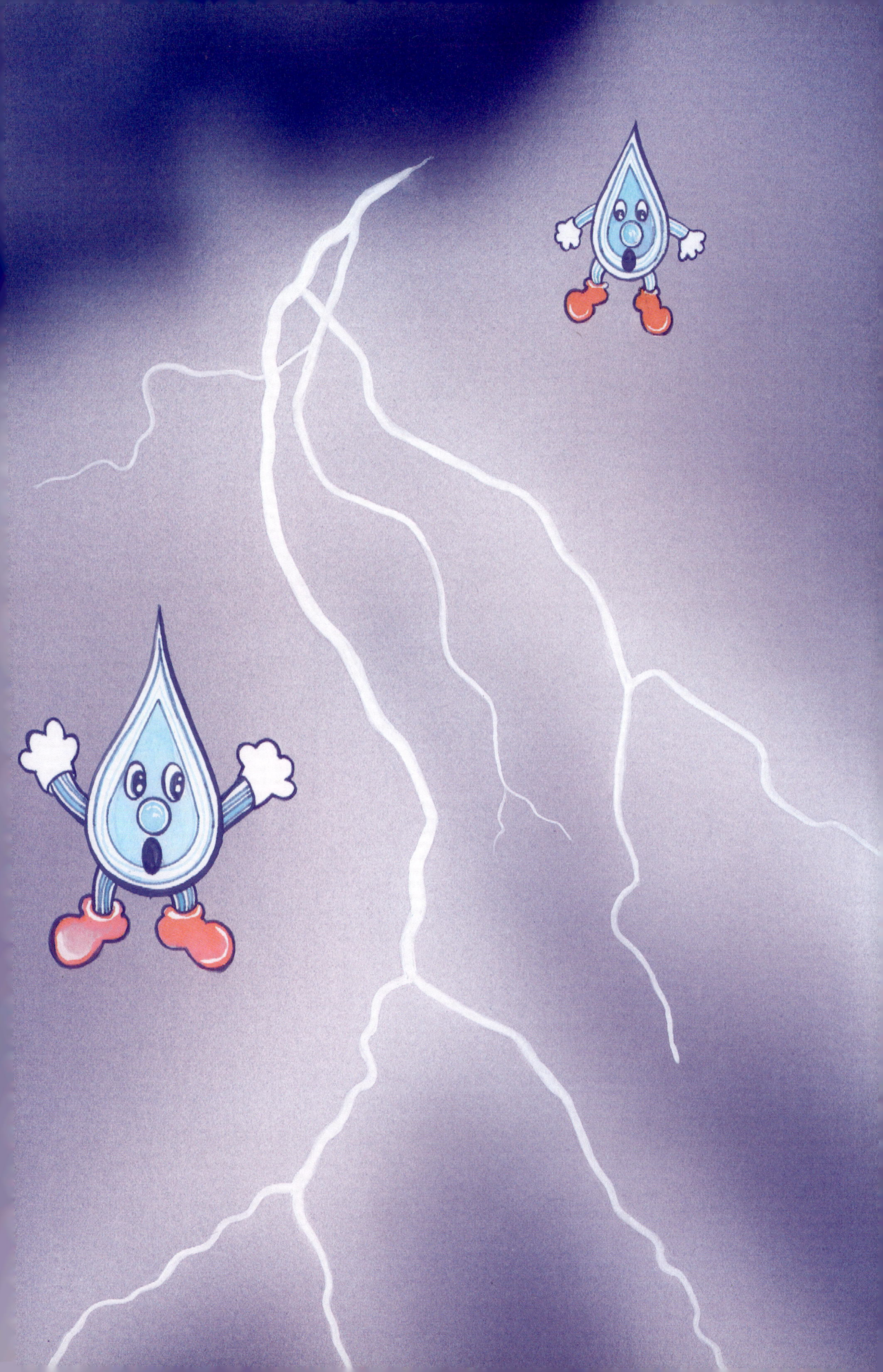

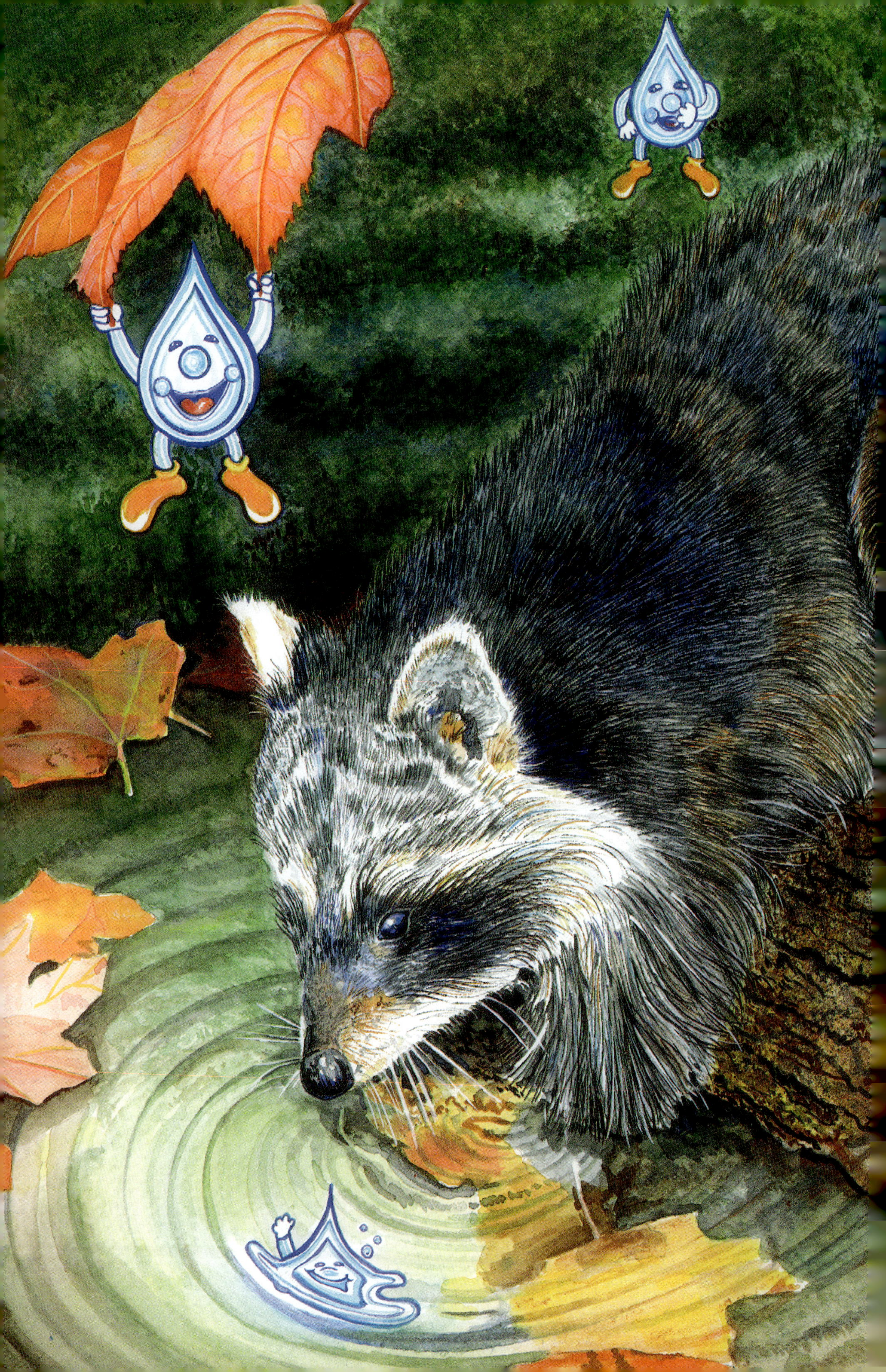

Fall is full of
so many colors to see.
Orange, red, and yellow
the leaves will soon be.

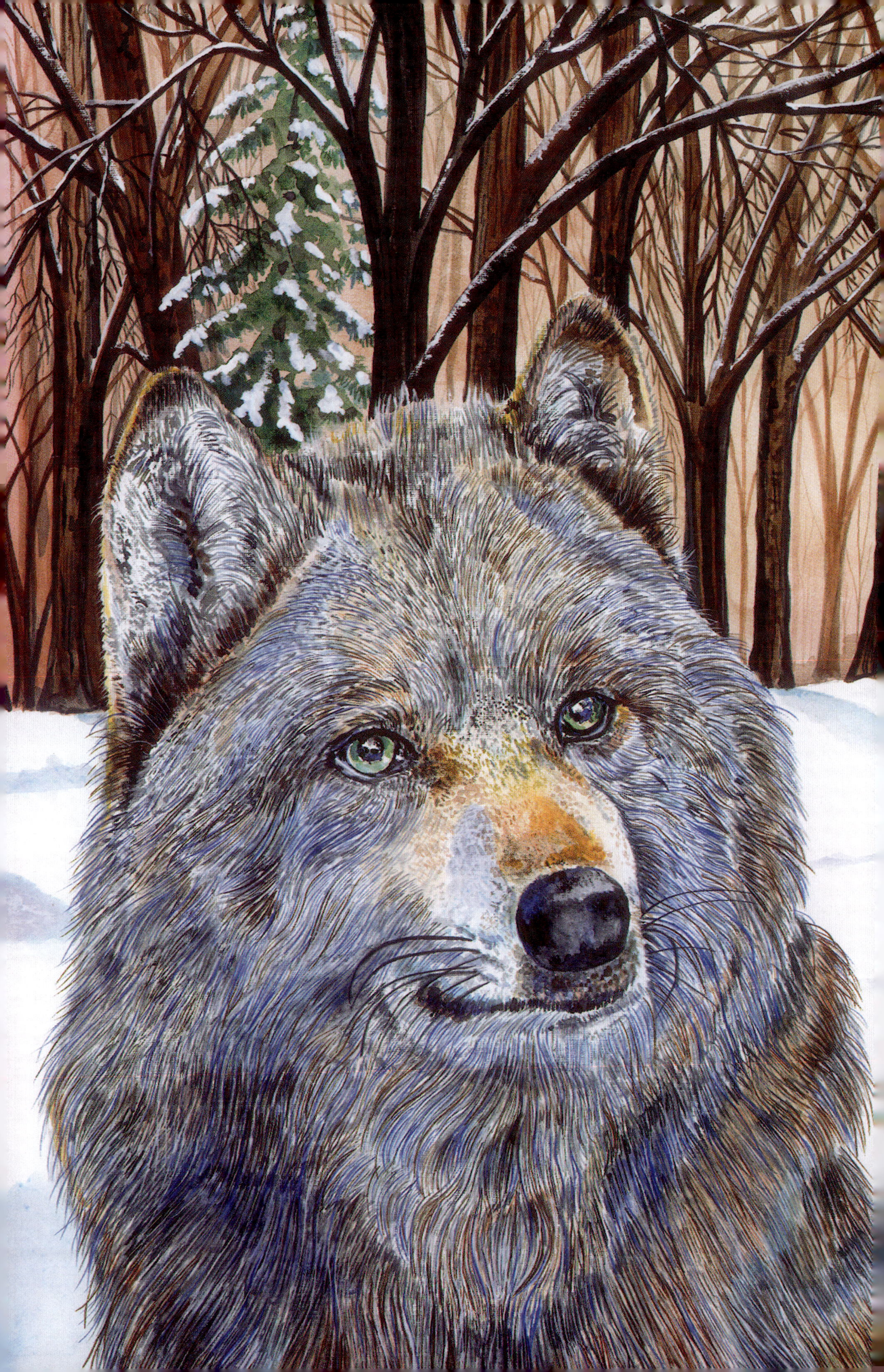

**The nights are getting colder.
Winter is almost here.
The rain turns into a snowflake,
no longer the shape of a tear.**

The raindrops' adventure
is almost done.
Slowly they will get pulled
back up by the Sun.

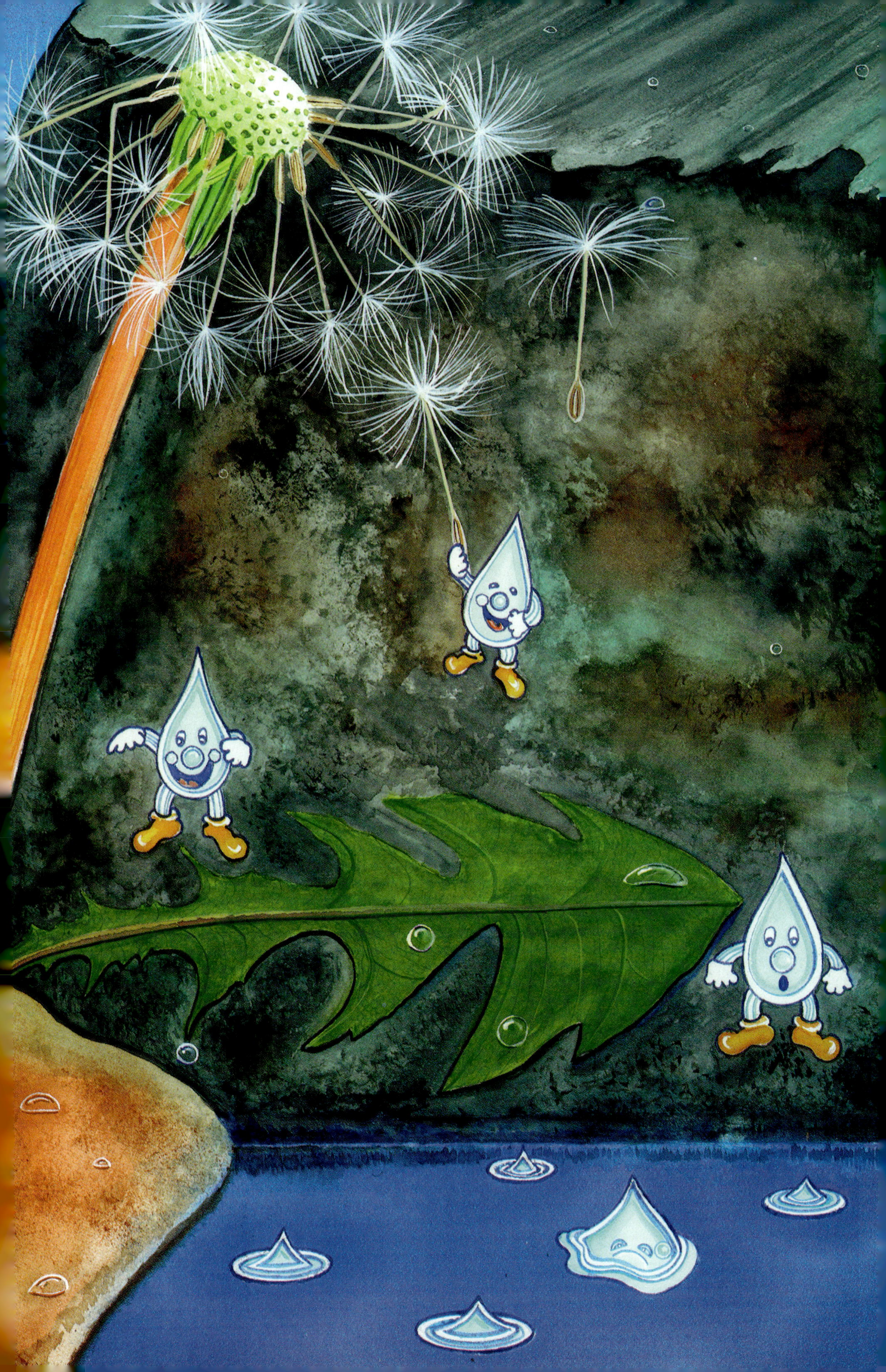

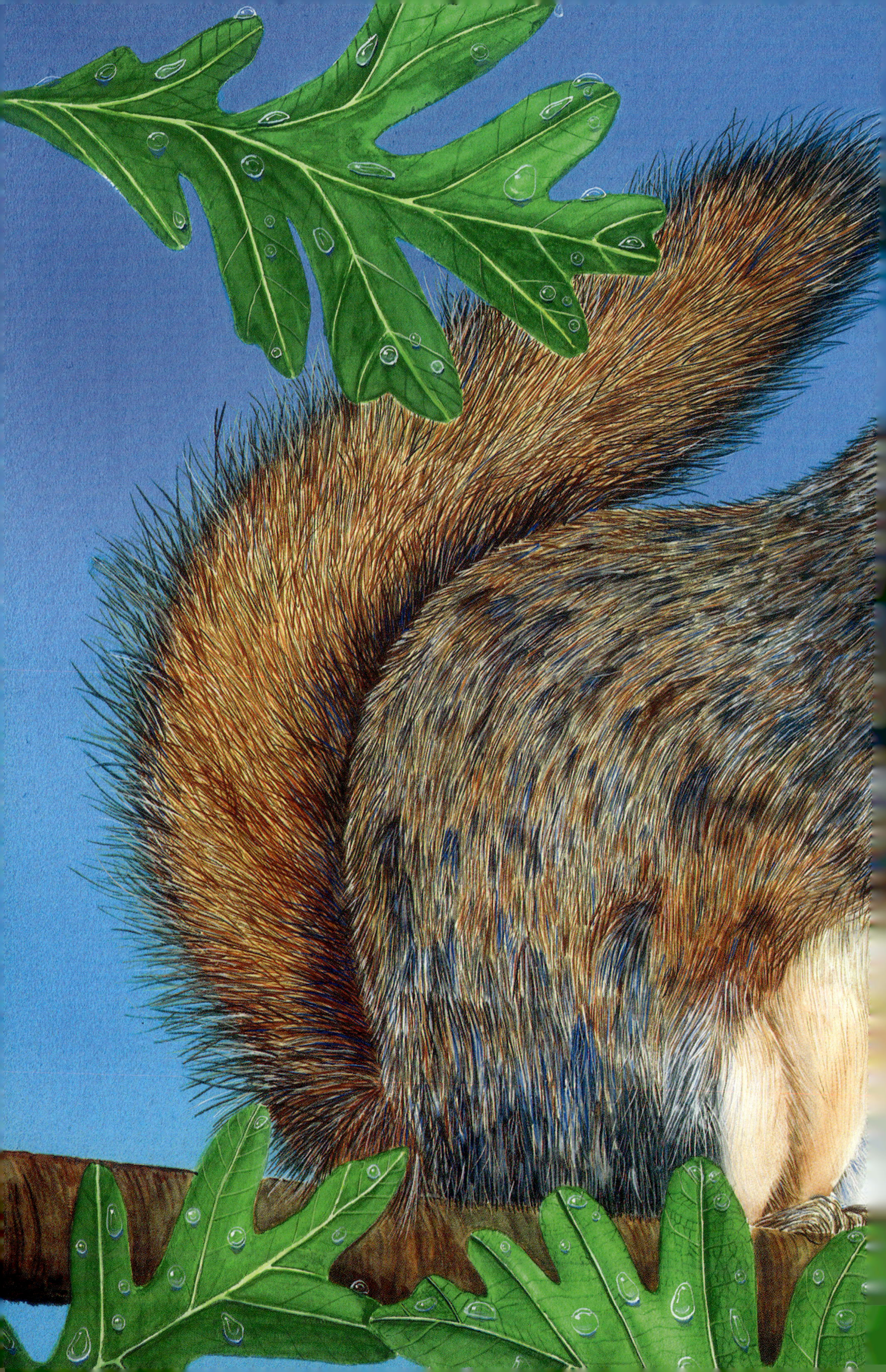

REMEMBER:
Evaporation and condensation,
they are the key.
That's how the wonderful
raindrops came to be!

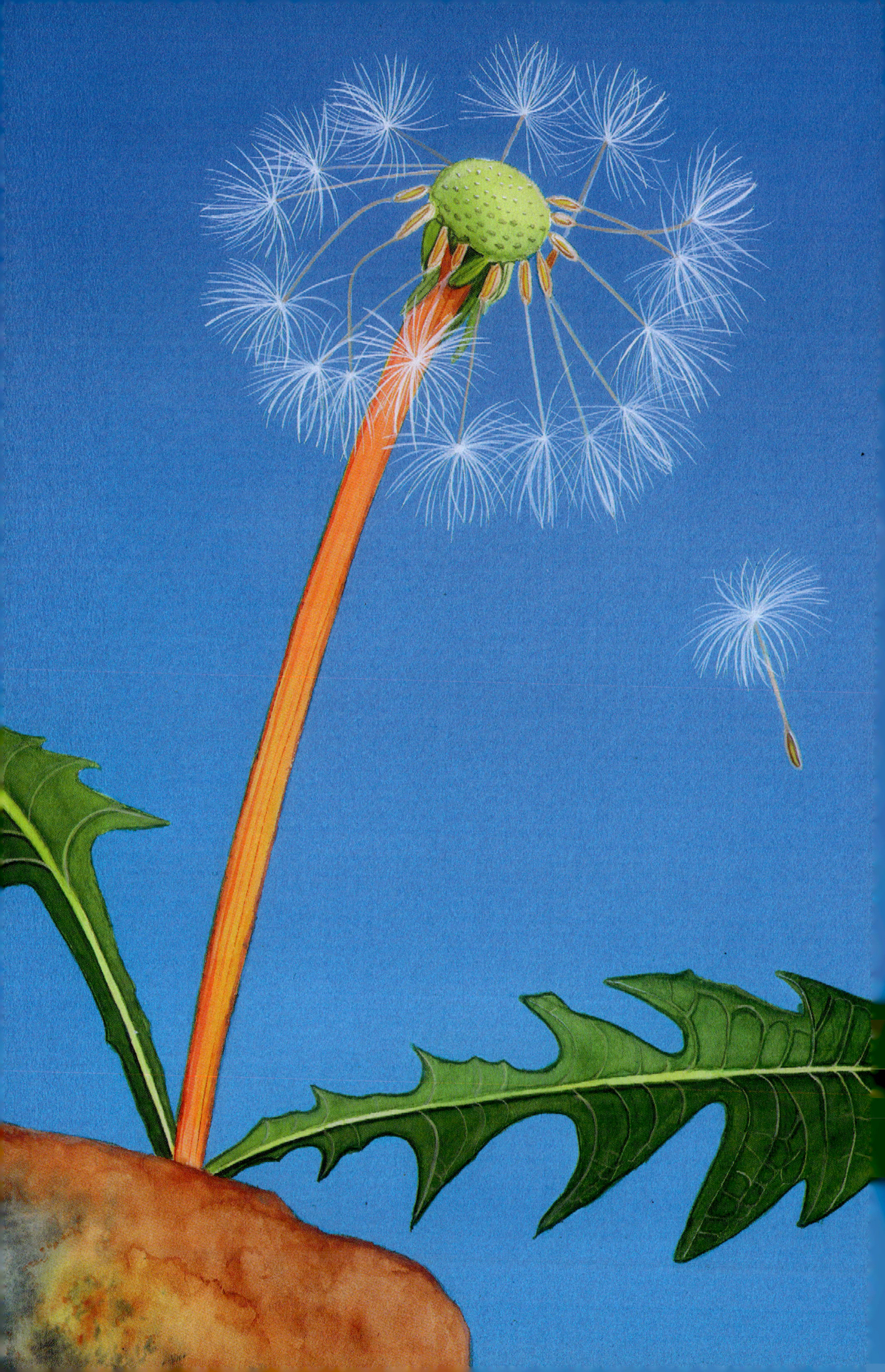

A dream is a wish
your heart needs.
So go make a wish
with dandelion seeds.
Gather your dreams
and let them fly.
With dreams your heart
will touch the sky.

Thank you,
Kimberly Kerr
1997

Oh No!
My Adventure
Is Almost
Done!
Don't Worry
Lil' Raindrop
We'll Soon Get
Pulled Back Up
To The Sun!

SEVEN COLORS IN A RAINBOW

There are seven colors in a Rainbow.
Do you know the order in which they go?
There are seven colors that you can see.
Seven colors are all there will be.
RED is the top color up by the sun.
ORANGE, then YELLOW, bright and fun.
GREEN is in the middle and he stands all alone.
BLUE, then INDIGO a color less known.
VIOLET is on the bottom, the seventh color you see.
Seven colors to the rainbow. Can you name them for me?

COPYRIGHT July 25, 1999 by Kimberly Michaud Kerr

FUN

INSIDE:
Book
Book Mark
Word Search
Word Find
Coloring
And More!

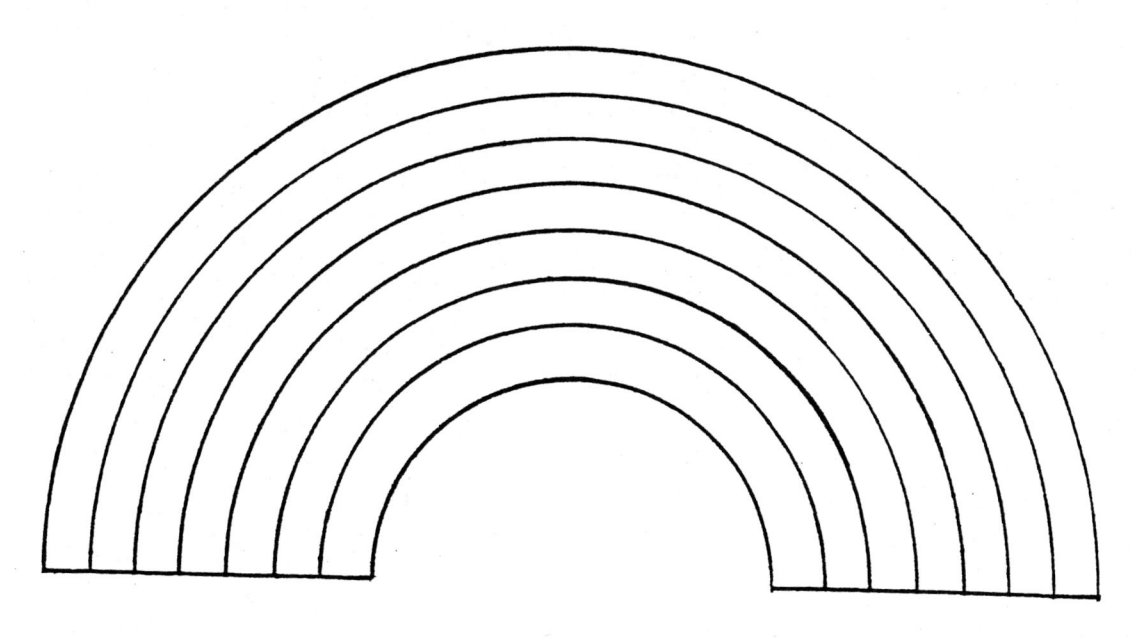

PACK

THE RAINDROPS' ADVENTURE

KIMBERLY KERR PRESS

COPYRIGHT 1999 by Kimberly Kerr

THE RAINDROPS' ADVENTURE

KIMBERLY KERR PRESS
COPYRIGHT 1999 by Kimberly Kerr

HOW MANY WORDS CAN YOU MAKE FROM THE WORD
RAINDROPS

DRAW A FUNNY FACE ON
THE RAINDROPS

THE RAINDROPS' ADVENTURE

KIMBERLY KERR PRESS
COPYRIGHT 1999 by Kimberly Kerr

Sun

THE RAINDROPS' ADVENTURE
KIMBERLY KERR PRESS
COPYRIGHT by Kimberly Kerr 1999

THE RAINDROPS' ADVENTURE

CAN YOU FIND THE WORDS?

L	O	B	S	N	O	W	F	L	A	K	E	S	U	N	S	H	I	N	E
I	R	F	U	T	E	A	Y	E	L	L	O	W	O	T	A	E	Z	T	O
G	R	E	L	O	R	A	I	V	T	O	R	E	A	Y	O	A	R	B	S
H	C	O	N	D	E	N	S	A	T	I	O	N	K	E	R	R	A	L	U
T	L	R	E	L	D	Z	I	P	R	L	M	S	P	E	C	T	R	U	M
N	O	A	A	O	E	T	M	O	L	E	C	U	L	E	S	I	A	E	M
I	U	N	D	W	B	E	D	R	E	A	M	N	M	I	K	V	I	S	E
N	D	G	K	I	M	L	W	A	T	E	R	A	I	N	E	W	N	K	R
G	R	E	E	N	I	O	H	T	F	W	A	D	V	E	R	U	R	E	S
R	S	R	P	T	K	I	M	I	A	O	I	N	D	I	G	O	X	R	E
E	U	E	B	E	E	V	I	O	L	R	N	A	I	N	T	R	Y	R	A
V	A	P	O	R	R	E	T	N	L	K	D	G	R	O	U	N	D	K	S
C	O	N	D	E	N	S	E	K	I	M	R	E	F	R	A	C	T	E	O
L	O	T	E	A	E	R	A	I	N	B	O	W	B	C	F	R	O	A	N
A	D	V	E	N	T	U	R	E	S	S	P	R	I	N	G	R	E	Z	S

ADVENTURE	LIGHTNING	SPECTRUM
BLUE	MOLECULES	SPRING
CLOUD	ORANGE	SUMMER
CONDENSATION	RAINBOW	SUN
CONDENSE	RAINDROP	SUNSHINE
EVAPORATION	RED	VAPOR
FALL	REFRACT	VIOLET
GREEN	SEASONS	WATER
GROUND	SKY	WINTER
INDIGO	SNOWFLAKE	YELLOW

KIMBERLY KERR PRESS
COPYRIGHT by Kimberly Kerr 1999

Do you know what these words mean?

THE RAINDROPS' ADVENTURE

Match the words by drawing a line to the correct definition

Evaporation   A mass of tiny drops of water or ice particles floating high in the sky.

Condensation   A flash of light in the sky caused by a discharge of electricity between clouds, or between a cloud and the earth's surface.

Molecules   A band of colors into which white light is separated by being passed through a prism, or by other means.

Refract   To change into vapor. To remove the liquid or moisture.

Spectrum   To change from a gas to a liquid or solid form.

Cloud   To bend from a straight course.

Lightning   The simplest structure made up of two or more atoms that are joined by a pair of shared electrons.

THE RAINDROPS' ADVENTURE
KIMBERLY KERR PRESS
COPYRIGHT 1999 by Kimberly Kerr

Let's catch some

RAINDROPS

This is a fun activity that children of all ages will enjoy!

MATERIALS:
Rain (or gently dripping water)
Bowl
2 ½ cups of Flour
1 cup of Salt
Fork
Plate
Flat Tray

1. Mix the flour and salt together in the bowl.

2. Spread the mixture onto the flat tray.

3. Set the tray outside in the rain for a minute and then bring it back indoors. If there's no rain, gently drip water from your hand over the mixture.

Let the rain-catching mixture set for a few hours. After that time the flour and salt mixture will have created a mold around the shape the water made on impact. The shapes can be carefully scooped out with a fork and set on a plate in a warm place to harden.

Compare the sizes and shapes of the drops.
Why does hitting the ground make them change?
Repeat this activity in different kinds of rain.

THE RAINDROPS' ADVENTURE
KIMBERLY KERR PRESS
COPYRIGHT by Kimberly Kerr 1999

THE RAINDROPS' ADVENTURE

START

TRY MY MAZE!

FINISH

COPYRIGHT 1999 by Kimberly Kerr

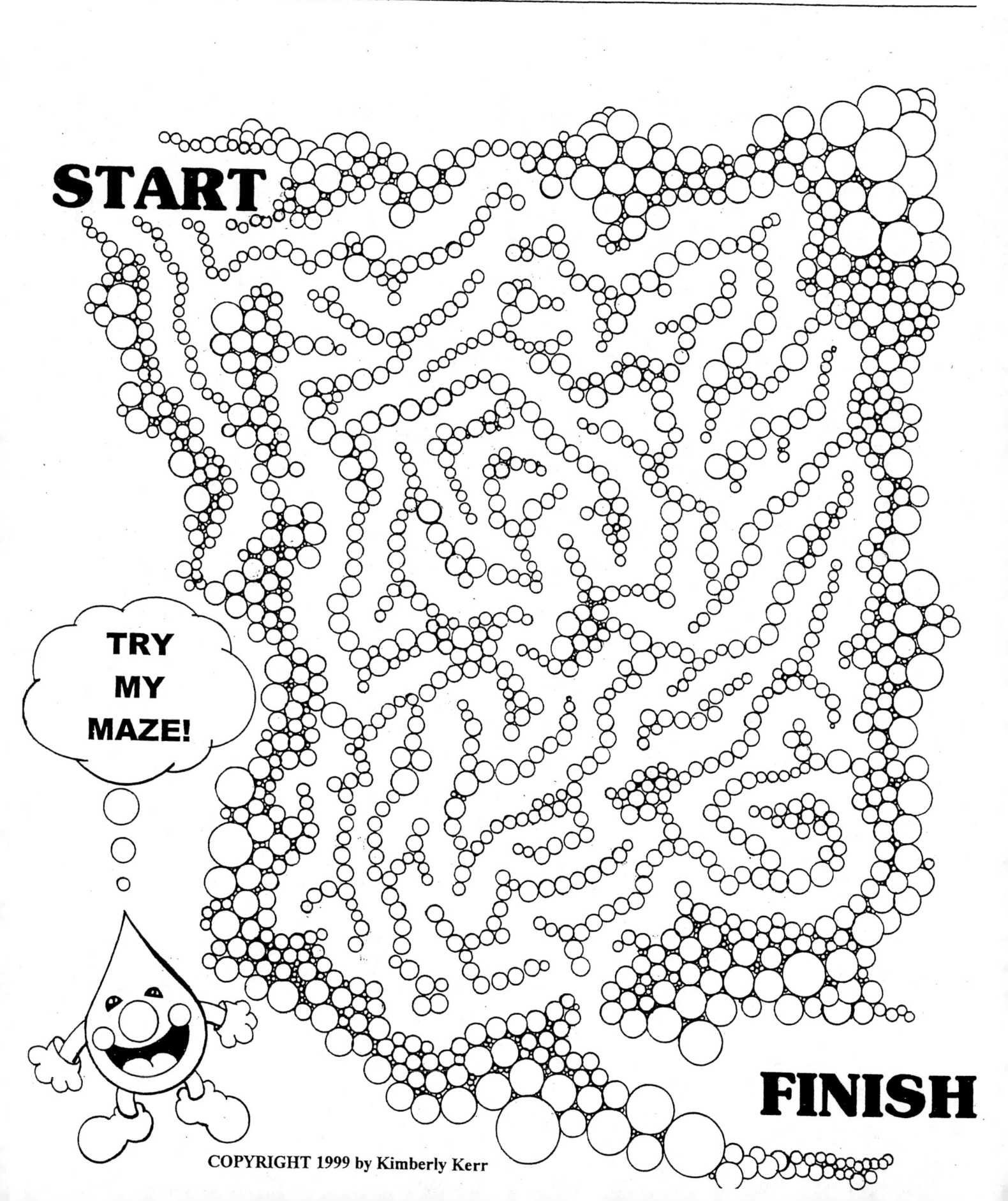

CREATING RAINBOWS
A mirror can create a visible rainbow!

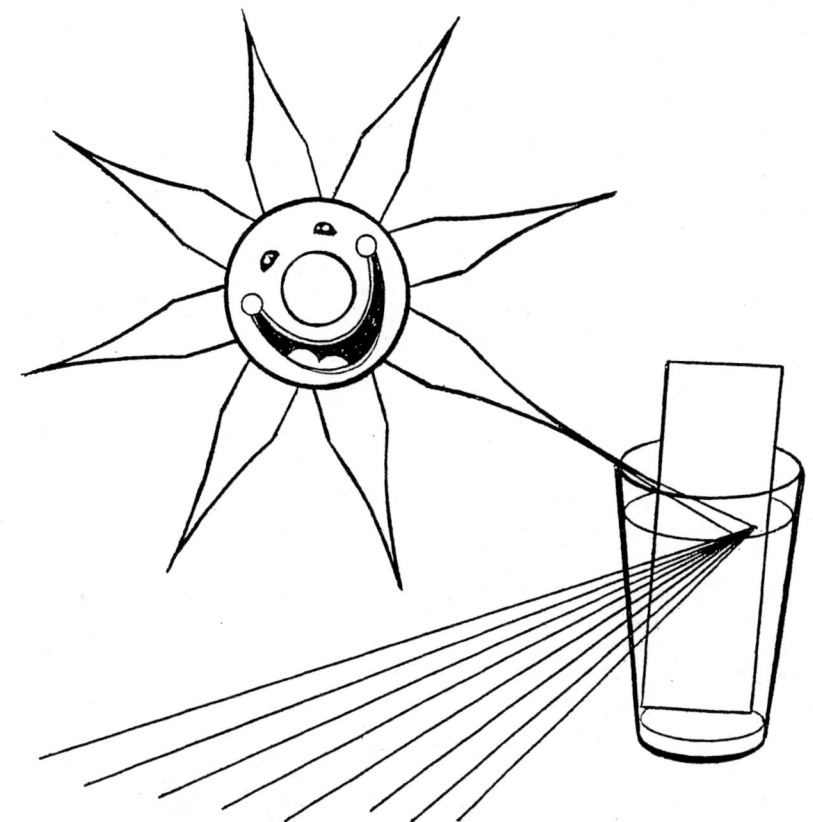

MATERIALS:
A small unbreakable mirror
A glass of water
ADULT SUPERVISION REQUIRED

Carefully place the glass of water in direct sunlight and then submerge a small mirror halfway in the water (half of the mirror is below and half is above the water line). By moving and rotating the mirror you will catch the sunlight, which will then be refracted through the water to create the rainbow colors. Observe light refraction while experimenting with light reflection in different parts of the room.

THE RAINDROPS' ADVENTURE

KIMBERLY KERR PRESS
COPYRIGHT 1999 by Kimberly Kerr

KIMBERLY KERR
PRESS

THANK-YOU FOR PURCHASING
THE RAINDROPS' ADVENTURE

ORDER ANOTHER COPY OF
THE RAINDROPS' ADVENTURE
SAVE $5.00

NAME_____

STREET_____

CITY, STATE, ZIP_____

To have your book autographed print the name below:

_____

The retail price of THE RAINDROPS' ADVENTURE is $14.95
With this coupon you will receive THE RAINDROPS' ADVENTURE at a reduced price of $9.95 plus 70 cents tax for Pennsylvania residents.
Please add $3.00 for shipping and handling.

PENNSYLVANIA RESIDENTS TOTAL IS $13.65
ALL OTHER STATES TOTAL IS $13.00

Please make checks payable to:
KIMBERLY KERR PRESS
P.O. BOX 63
SEWICKLEY, PA 15143

ENCLOSED IS $_____ FOR _____ COPIES

THANK-YOU! ENJOY YOUR BOOK!

Kimberly Kerr ~ Owner

Phone/Fax (412) 741-3656 • P. O. Box 63 • Sewickley, PA 15143